The New Heavens

by George Ellery Hale

TO MY WIFE

PREFACE

Fourteen years ago, in a book entitled "The Study of Stellar Evolution" (University of Chicago Press, 1908), I attempted to give in untechnical language an account of some modern methods of astrophysical research. This book is now out of print, and the rapid progress of science has left it completely out of date. As I have found no opportunity to prepare a new edition, or to write another book of similar purpose, I have adopted the simpler expedient of contributing occasional articles on recent developments to Scribner's Magazine, three of which are included in the present volume.

I am chiefly indebted, for the illustrations, to the Mount Wilson Observatory and the present and former members of its staff whose names appear in the captions. Special thanks are due to Mr. Ferdinand Ellerman, who made all of the photographs of the observatory buildings and instruments, and prepared all material for reproduction. The cut of the original Cavendish apparatus is copied from the Philosophical Transactions for 1798 with the kind permission of the Royal Society, and I am also indebted to the Royal Society and to Professor Fowler and Father Cortie for the privilege of reproducing from the Proceedings two illustrations of their spectroscopic results.

G. E. H.

January, 1922.

CONTENTS

CHAPTER I

THE NEW HEAVENS

Go out under the open sky, on a clear and moon-less night, and try to count the stars. If your station lies well beyond the glare of cities, which is often strong enough to conceal all but the brighter objects, you will find the task a difficult one. Ranging through the six magnitudes of the Greek astronomers, from the brilliant Sirius to the faintest perceptible points of light, the stars are scattered in great profusion over the celestial vault. Their number seems limitless, yet actual count will show that the eye has been deceived. In a survey of the entire heavens, from pole to pole, it would not be possible to detect more than from six to seven thousand stars with the naked eye. From a single viewpoint, even with the keenest vision, only two or three thousand can be seen. So many of these are at the limit of visibility that Ptolemy's "Almagest," a catalogue of all the stars whose places were measured with the simple instruments of the Greek astronomers, contains only 1,022 stars.

Back of Ptolemy, through the speculations of the Greek philosophers, the mysteries of the Egyptian sun-god, and the observations of the ancient Chaldeans, the rich and varied traditions of astronomy stretch far away into a shadowy past. All peoples, in the first stirrings of their intellectual youth, drawn by the nightly splendor of the skies and the ceaseless motions of the planets, have set up some system of the heavens, in which the sense of wonder and the desire for knowledge were no less concerned than the practical necessities of life. The measurement of time and the needs of navigation have always stimulated astronomical research, but the intellectual demand has been keen from the first. Hipparchus and the Greek astronomers of the Alexandrian school, shaking off the vagaries of magic and divination, placed astronomy on a scientific basis, though the reaction of the Middle Ages caused even such a great astronomer as Tycho Brahe himself to revert for a time to the practice of astrology.

EARLY INSTRUMENTS

The transparent sky of Egypt, rarely obscured by clouds, greatly favored Ptolemy's observations. Here was prepared his great star catalogue, based upon the earlier observations of Hipparchus, and destined to remain alone in its field for more than twelve centuries, until Ulugh Bey, Prince of Samarcand, repeated the work of his Greek predecessor. Throughout this period the stars were looked upon mainly as points of reference for the observation of planetary motions, and the instruments of observation underwent little change. The astrolabe, which consists of a circle divided into degrees, with a rotating diametral arm for sighting purposes, embodies their essential principle. In its simple form, the astrolabe was suspended in a vertical plane, and the stars were observed by bringing the sights on the movable diameter to bear upon them. Their altitude was then read off on the circle. Ultimately, the circle of the astrolabe, mounted with one of its diameters parallel to the earth's axis, became the armillary sphere, the precursor of our modern equatorial telescope. Great stone quadrants fixed in the meridian were also employed from very early times. Out of such furnishings, little modified by the lapse of centuries, was provided the elaborate instrumental equipment of Uranibourg, the great observatory built by Tycho Brahe on the Danish island of Huen in 1576. In this "City of the Heavens," still dependent solely upon the unaided eye as a collector of starlight, Tycho made those invaluable observations that enabled Kepler to deduce the true laws of planetary motion. But after all these centuries the sidereal world embraced no objects, barring an occasional comet or temporary star, that lay beyond the vision of the earliest astronomers. The conceptions of the stellar universe, except those that ignored the solid ground of observation, were limited by the small aperture of the human eye. But the dawn of another age was at hand.

The dominance of the sun as the central body of the solar system, recognized by Aristarchus of Samos nearly three centuries before the Christian era, but subsequently denied under the authority of Ptolemy and the teachings of the Church, was reaffirmed by the Polish monk Copernicus in 1543. Kepler's laws of the motions of the planets, showing them to revolve in ellipses instead of circles, removed the last defect of the Copernican system, and left no room for its rejection. But both the world and the Church clung to tradition, and some visible demonstration was urgently needed. This was supplied by Galileo through his invention of the telescope.

The 60-foot tower on the extreme left, which is at the edge of a precipitous ca 駧 n 1,500 feet deep, is the vertical telescope of the Smithsonian Astrophysical Observatory. Above it are the "Monastery" and other buildings used as quarters by the astronomers of the Mount Wilson Observatory while at work on the mountain. (The offices, computing-rooms, laboratories, and shops are in Pasadena.) Following the ridge, we come successively to the dome of the 10-inch photographic telescope, the power-house, laboratory, Snow horizontal telescope, 60-foot-tower telescope, and 150-foot-tower telescope, these last three used for the study of the sun. The dome of the 60-inch reflecting telescope is just below the 150-foot tower, while that of the 100-inch telescope is farther to the right. The altitude of Mount Wilson is about 5,900 feet.]

The crystalline lens of the human eye, limited by the iris to a maximum opening about one-quarter of an inch in diameter, was the only collector of starlight available to the Greek and Arabian astronomers. Galileo's telescope, which in 1610 suddenly pushed out the boundaries of the known stellar universe and brought many thousands of stars into range, had a lens about 2-1/4 inches in diameter. The area of this lens, proportional to the square of its diameter, was about eighty-one times that of the pupil of the eye. This great increase in the amount of light collected should bring to view stars down to magnitude 10.5, of which nearly half a million are known to exist.

It is not too much to say that Galileo's telescope revolutionized human thought. Turned to the moon, it revealed mountains, plains, and valleys, while the sun, previously supposed immaculate in its perfection, was seen to be blemished with dark spots changing from day to day. Jupiter, shown to be accompanied by four encircling satellites, afforded a picture in miniature of the solar system, and strongly supported the Copernican view of its organization, which was conclusively demonstrated by Galileo's discovery of the changing phases of Venus and the variation of its apparent diameter during its revolution about the sun. Galileo's proof of the Copernican theory marked the downfall of medievalism and established astronomy on a firm foundation. But while his telescope multiplied a hundredfold the number of visible stars, more than a century elapsed before the true possibilities of sidereal astronomy were perceived.

STRUCTURE OF THE UNIVERSE

Sir William Herschel was the first astronomer to make a serious attack upon the problem of the structure of the stellar universe. In his first memoir on the "Construction of the Heavens," read before the Royal Society in 1784, he wrote as follows:

"Hitherto the sidereal heavens have, not inadequately for the purpose designed, been represented by the concave surface of a sphere in the centre of which the eye of an observer might be supposed to be placed.... In future we shall look upon those regions into which we may now penetrate by means of such large telescopes, as a naturalist regards a rich extent of ground or chain of mountains containing strata variously inclined and directed as well as consisting of very different materials."

On turning his 18-inch reflecting telescope to a part of the Milky Way in Orion, he found its whitish appearance to be completely resolved into small stars, not separately seen with his former telescopes. "The glorious multitude of stars of all possible sizes that presented themselves here to my view are truly astonishing; but as the dazzling brightness of glittering stars may easily mislead us so far as to estimate their number greater than it really is, I endeavored to ascertain this point by counting many fields, and computing from a mean of them, what a certain given portion of the Milky Way might contain." By this means, applied not only to the Milky Way but to all parts of the heavens, Herschel determined the approximate number and distribution of all the stars within reach of his instrument.

By comparing many hundred gauges or counts of stars visible in a field of about one-quarter of the area of the moon, Herschel found that the average number of stars increased toward the great circle which most nearly conforms with the course of the Milky Way. Ninety degrees from this plane, at the pole of the Milky Way, only four stars, on the average, were seen in the field of the telescope. In approaching the Milky Way this number increased slowly at first, and then more and more rapidly, until it rose to an average of 122 stars per field.

These observations were made in the northern hemisphere, and subsequently Sir John Herschel, using his father's telescope at the Cape of Good Hope, found an almost exactly similar increase of apparent star density for the southern hemisphere. According to his estimates, the total number of stars in both hemispheres that could be seen distinctly enough to be counted in this telescope would probably be about five and one-half millions.

The Herschels concluded that "the stars of our firmament, instead of being scattered in all directions indifferently through space, form a stratum of which the thickness is small, in comparison with its length and breadth; and in which the earth occupies a place somewhere about the middle of its thickness, between the point where it subdivides into two principal lamin?inclined at a small angle to each other." This view does not differ essentially from our modern conception of the form of the Galaxy; but as the Herschels were unable to see stars fainter than the fifteenth magnitude, it is evident that their conclusions apply only to a restricted region surrounding the solar system, in the midst of the enormously extended sidereal universe which modern instruments have brought within our range.

MODERN METHODS

The remarkable progress of modern astronomy is mainly due to two great instrumental advances: the rise and development of the photographic telescope, and the application of the spectroscope to the study of celestial objects. These new and powerful instruments, supplemented by many accessories which have completely revolutionized observatory equipment, have not only revealed a vastly greater number of stars and nebul? they have also rendered feasible observations of a type formerly regarded as impossible. The chemical analysis of a faint star is now so easy that it can be accomplished in a very short time--as quickly, in fact, as an equally complex substance can be analyzed in the laboratory. The spectroscope also measures a star's velocity, the pressure at different levels in its atmosphere, its approximate temperature, and now, by a new and ingenious method, its distance from the earth. It determines the velocity of rotation of the sun and of nebul? the existence and periods of orbital revolution of binary stars too close to be separated by any telescope, the presence of magnetic fields in sunspots, and the fact that the entire sun, like the earth, is a magnet.

Such new possibilities, with many others resulting from the application of physical methods of the most diverse character, have greatly enlarged the astronomer's outlook. He may now attack two great problems: (1) The structure of the universe and the motions of its constituent bodies, and (2) the evolution of the stars: their nature, origin, growth, and decline. These two problems are intimately related and must be studied as one.[*]

[Footnote *: A third great problem open to the astronomer, the study of the constitution of matter, is described in Chapter III.]

If space permitted, it would be interesting to survey the progress already accomplished by modern methods of astronomical research. Hundreds of millions of stars have been photographed, and the boundaries of the stellar universe have been pushed far into space, but have not been attained. Globular star clusters, containing tens of thousands of stars, are on so great a scale (according to Shapley) that light, travelling at the rate of 186,000 miles per second, may take 500 years to cross one of them, while the most distant of these objects may be more than 200,000 light-years from the earth. The spiral nebul? more than a million in number, are vast whirling masses in process of development, but we are not yet certain whether they should be regarded as "island universes" or as subordinate to the stellar system which includes our minute group of sun and planets, the great star clouds of the Milky Way, and the distant globular star clusters.

These few particulars may give a slight conception of the scale of the known universe, but a word must be added regarding some of its most striking phenomena. The great majority of the stars whose motions have been determined belong to one or the other of two great star streams, but the part played by these streams in the sidereal system as a whole is still obscure. The stars have been grouped in classes, presumably in the order of their evolutional development, as they pass from the early state of gaseous masses, of low density, through the successive stages resulting from loss of heat by radiation and increased density due to shrinkage. Strangely enough, their velocities in space show a corresponding change, increasing as they grow older or perhaps depending upon their mass.

It is impossible within these limits to do more than to give some indication of the scope of the new astronomy. Enough has been said, however, to assist in appreciating the increased opportunity for investigation, and the nature of the heavy demands made upon the modern observatory. But before passing on to describe one of the latest additions to the astronomer's instrumental equipment, a word should be added regarding the chief classes of telescopes.

REFRACTORS AND REFLECTORS

Astronomical telescopes are of two types: refractors and reflectors. A refracting telescope consists of an object-glass composed of two or more lenses, mounted at the upper end of a tube, which is pointed at the celestial object. The light, after passing through the lenses, is brought to a focus at the lower end of the tube, where the image is examined visually with an eyepiece, or photographed upon a sensitive plate. The largest instruments of this type are the 36-inch Lick telescope and the 40-inch refractor of the Yerkes Observatory.

Reflecting telescopes, which are particularly adapted for photographic work, though also excellent for visual observations, are very differently constructed. No lens is used. The telescope tube is usually built in skeleton form, open at its upper end, and with a large concave mirror supported at its base. This mirror serves in place of a lens. Its upper surface is paraboloidal in shape, as a spherical surface will not unite in a sharp focus the rays coming from a distant object. The light passes through no glass--a great advantage, especially for photography, as the absorption in lenses cuts out much of the blue and violet light, to which photographic plates are most sensitive. The reflection occurs on the upper surface of the mirror, which is covered with a coat of pure silver, renewed several times a year and always kept highly burnished. Silvered glass is better than metals or other substances for telescope mirrors, chiefly because of the perfection with which glass can be ground and polished, and the ease of renewing its silvered surface when tarnished.

The great reflectors of Herschel and Lord Rosse, which were provided with mirrors of speculum metal, were far inferior to much smaller telescopes of the present day. With these instruments the star images were watched as they were carried through the field of view by the earth's rotation, or kept roughly in place by moving the telescope with ropes or chains. Photographic plates, which reveal invisible stars and nebul?when exposed for hours in modern instruments, were not then available. In any case they could not have been used, in the absence of the perfect mechanism required to keep the star images accurately fixed in place upon the sensitive film.

It would be interesting to trace the long contest for supremacy between refracting and reflecting telescopes, each of which, at certain stages in its development, appeared to be unrivalled. In modern observatories both types are used, each for the purpose for which it is best adapted. For the photography of nebul?and the study of the fainter stars, the reflector has special advantages, illustrated by the work of such instruments as the Crossley and Mills reflectors of the Lick Observatory; the great 72-inch reflector, recently brought into effective service at the Dominion Observatory in Canada; and the 60-inch and 100-inch reflectors of the Mount Wilson Observatory.

The unaided eye, with an available area of one-twentieth of a square inch, permits us to see stars of the sixth magnitude. Herschel's 18-inch reflector, with an area 5,000 times as great, rendered visible stars of the fifteenth magnitude. The 60-inch reflector, with an area 57,600 times that of the eye, reveals stars of the eighteenth magnitude, while to reach stars of about the twentieth magnitude, photographic exposures of four or five hours suffice with this instrument.

Every gain of a magnitude means a great gain in the number of stars rendered visible. Stars of the second magnitude are 3.4 times as numerous as those of the first, those of the eighth magnitude are three times as numerous as those of the seventh, while the sixteenth magnitude stars are only 1.7 as numerous as those of the fifteenth magnitude. This steadily decreasing ratio is probably due to an actual thinning out of the stars toward the boundaries of the stellar universe, as the most exhaustive tests have failed to give any evidence of absorption of light in its passage through space. But in spite of this decrease, the gain of a single additional magnitude may mean the addition of many millions of stars to the total of those already shown by the 60-inch reflector. Here is one of the chief sources of interest in the possibilities of a 100-inch reflecting telescope.

100-INCH TELESCOPE

In 1906 the late John D. Hooker, of Los Angeles, gave the Carnegie Institution of Washington a sum sufficient to construct a telescope mirror 100 inches in diameter, and thus large enough to collect 160,000 times the light received by the eye. (Fig. 10.) The casting and annealing of a suitable glass disk, 101 inches in diameter and 13 inches thick, weighing four and one-half tons, was a most difficult operation, finally accomplished by a great French glass company at their factory in the Forest of St. Gobain. A special optical laboratory was erected at the Pasadena headquarters of the Mount Wilson Observatory, and here the long task of grinding, figuring, and testing the mirror was successfully carried out by the observatory opticians. This operation, which is one of great delicacy, required years for its completion. Meanwhile the building, dome, and mounting for the telescope were designed by members of the observatory staff, and the working drawings were prepared. An opportune addition by Mr. Carnegie to the endowment of the Carnegie Institution of Washington, of which the observatory is a branch, permitted the necessary appropriations to be made for the completion and erection of the telescope. Though delayed by the war, during which the mechanical and optical facilities of the observatory shops were utilized for military and naval purposes, the telescope is now in regular use on Mount Wilson.

The instrument is mounted on a massive pier of reinforced concrete, 33 feet high and 52 feet in diameter at the top. A solid wall extends south from this pier a distance of 50 feet, on the west side of which a very powerful spectrograph, for photographing the spectra of the brightest stars, will be mounted. Within the pier are a photographic dark room, a room for silvering the large mirror (which can be lowered into the pier), and the clock-room, where stands the powerful driving-clock, with which the telescope is caused to follow the apparent motion of the stars. (Fig. 11.)

The telescope mounting is of the English type, in which the telescope tube is supported by the declination trunnions between the arms of the polar axis, built in the form of a rectangular yoke carried by bearings on massive pedestals to the north and south. These bearings must be aligned exactly parallel to the axis of the earth, and must support the polar axis so freely that it can be rotated with perfect precision by the driving-clock, which turns a worm-wheel 17 feet in diameter, clamped to the lower end of the axis. As this motion must be sufficiently uniform to counteract exactly the rotation of the earth on its axis, and thus to maintain the star images accurately in position in the field of view, the greatest care had to be taken in the construction of the driving-clock and in the spacing and cutting of the teeth in the large worm-wheel. Here, as in the case of all of the more refined parts of the instrument, the work was done by skilled machinists in the observatory shops in Pasadena or on Mount Wilson after the assembling of the telescope. The massive sections of the instrument, some of which weigh as much as ten tons each, were constructed at Quincy, Mass., where machinery sufficiently large to build battleships was available. They were then shipped to California, and transported to the summit of Mount Wilson over a road built for this purpose by the construction division of the observatory, which also built the pier on which the telescope stands, and erected the steel building and dome that cover it.

The parts of the telescope which are moved by the driving-clock weigh about 100 tons, and it was necessary to provide means of reducing the great friction on the bearings of the polar axis. To accomplish this, large hollow steel cylinders, floating in mercury held in cast-iron tanks, were provided at the upper and lower ends of the polar axis. Almost the entire weight of the instrument is thus floated in mercury, and in this way the friction is so greatly reduced that the driving-clock moves the instrument with perfect ease and smoothness.

The 100-inch mirror rests at the bottom of the telescope tube on a special support system, so designed as to prevent any bending of the glass under its own weight. Electric motors, forty in number, are provided to move the telescope rapidly or slowly in right ascension (east or west) and in declination (north or south), for focussing the mirrors, and for many other purposes. They are also used for rotating the dome, 100 feet in diameter, under which the telescope is mounted, and for opening the shutter, 20 feet wide, through which the observations are made.

A telescope of this kind can be used in several different ways. The 100-inch mirror has a focal length of about 42 feet, and in one of the arrangements of the instrument, the photographic plate is mounted at the centre of the telescope tube near its upper end, where it receives directly the image formed by the large mirror. In another arrangement, a silvered glass mirror, with plane surface, is supported near the upper end of the tube at an angle of 45? so as to form the image at the side of the tube, where the photographic plate can be placed. In this case, the observer stands on a platform, which is moved up and down by electric motors in front of the opening in the dome through which the observations are made.

Other arrangements of the telescope, for which auxiliary convex mirrors carried near the upper end of the tube are required, permit the image to be photographed at the side of the tube near its lower end, either with or without a spectrograph; or with a very powerful spectrograph mounted within a constant-temperature chamber south of the telescope pier. In this last case, the light of a star is so reflected by auxiliary mirrors that it passes down through a hole in the south end of the polar axis and brings the star to a focus on the slit of the fixed spectrograph.

ATMOSPHERIC LIMITATIONS

The huge dimensions of such a powerful engine of research as the Hooker telescope are not in themselves a source of satisfaction to the astronomer, for they involve a decided increase in the labor of observation and entail very heavy expense, justifiable only in case important results, beyond the reach of other instruments, can be secured. The construction of a telescope of these dimensions was necessarily an experiment, for it was by no means certain, after the optical and mechanical difficulties had been overcome, that even the favorable atmosphere of California would be sufficiently tranquil to permit sharply defined celestial images to be obtained with so large an aperture. It is therefore important to learn what the telescope will actually accomplish under customary observing conditions.

Fortunately we are able to measure the performance of the instrument with certainty. Close beside it on Mount Wilson stands the 60-inch reflector, of similar type, erected in 1908. The two telescopes can thus be rigorously compared under identical atmospheric conditions.

The large mirror of the 100-inch telescope has an area about 2.8 times that of the 60-inch, and therefore receives nearly three times as much light from a star. Under atmospheric conditions perfect enough to allow all of this light to be concentrated in a point, it should be capable of recording on a photographic plate, with a given exposure, stars about one magnitude fainter than the faintest stars within reach of the 60-inch. The increased focal length, permitting such objects as the moon to be photographed on a larger scale, should also reveal smaller details of structure and render possible higher accuracy of measurement. Finally, the greater theoretical resolving power of the larger aperture, providing it can be utilized, should permit the separation of the members of close double stars beyond the range of the smaller instrument.

CRITICAL TESTS

The many tests already made indicate that the advantages expected of the new telescope will be realized in practice. The increased light-gathering power will mean the addition of many millions of stars to those already known. Spectroscopic observations now in regular progress have carried the range of these investigations far beyond the possibilities of the 60-inch telescope. A great class of red stars, for example, almost all the members of which were inaccessible to the 60-inch, are now being made the subject of special study. And in other fields of research equal advantages have been gained.

The increase in the scale of the images over those given by the 60-inch telescope is illustrated by two photographs of the Ring Nebula in Lyra, reproduced in Fig. 18. The Great Nebula in Orion, photographed with the 100-inch telescope with a comparatively short exposure, sufficient to bring out the brighter regions, is reproduced in Fig. 2. It is interesting to compare this picture with the small-scale image of the same nebula shown in Fig. 1.

The ring-like formations are the so-called craters, most of them far larger than anything similar on the earth. That in the lower left corner with an isolated mountain in the centre is Albategnius, sixty-four miles in diameter. Peaks in the ring rise to a height of fifteen thousand feet above the central plain. Note the long sunset shadows cast by the mountains on the left. The level region below on the right is an extensive plain, the Mare Nubium.]

The mountains above and to the left are the lunar Apennines; those on the left just below the centre are the Alps. Both ranges include peaks from fifteen thousand to twenty thousand feet in height. In the upper right corner is Copernicus, about fifty miles in diameter. The largest of the conspicuous group of three just below the Apennines is Archimedes and at the lower end of the Alps is Plato. Note the long sunset shadows cast by the isolated peaks on the left. The central portion of the picture is a vast plain, the Mare Imbrium.]

The sharpness of the images given by the new telescope may be illustrated by some recent photographs of the moon, obtained with an equivalent focal length of 134 feet. In Fig. 15 is shown a rugged region of the moon, containing many ring-like mountains or craters. Fig. 16 shows the great arc of the lunar Apennines (above) and the Alps (below), to the left of the broad plain of the Mare Imbrium. The starlike points along the moon's terminator, which separates the dark area from the region upon which the sun (on the right) shines, are the mountain peaks, about to disappear at sunset. The long shadows cast by the mountains just within the illuminated area are plainly seen. Some of the peaks of the lunar Apennines attain a height of 20,000 feet.

In less powerful telescopes the stars at the centre of the great globular clusters are so closely crowded together that they cannot be studied separately with the spectrograph. Moreover, most of them are much too faint for examination with this instrument. At the 134-foot focus the 100-inch telescope gives a large-scale image of such clusters, and permits the spectra of stars as faint as the fifteenth magnitude to be separately photographed.

CLOSE DOUBLE STARS

A remarkable use of the 100-inch telescope, which permits its full theoretical resolving power to be not merely attained but to be doubled, has been made possible by the first application of Michelson's interference method to the measurement of very close double stars. When employing this, the 100-inch mirror is completely covered, except for two slits. Beams of light from a star, entering by the slits, unite at the focus of the telescope, where the image is examined by an eyepiece magnifying about five thousand diameters. Across the enlarged star image a series of fine, sharp fringes is seen, even when the atmospheric conditions are poor. If the star is single the fringes remain visible, whatever the distance between the slits. But in the case of a star like Capella, previously inferred to be double from the periodic displacement of the lines in its spectrum, but with components too close together to be distinguished separately, the fringes behave differently. As the slits are moved apart a point is reached where the fringes completely disappear, only to reappear as the separation is continued. This effect is obtained when the slits are at right angles to the line joining the two stars of the pair, found by this method to be 0.0418 of a second of arc apart (on December 30, 1919). Subsequent measures, of far greater precision than those obtainable by other methods in the case of easily separated double stars, show the rapid orbital motion of the components of the system. This device will be applied to other close binaries, hitherto beyond the reach of measurement.

Without entering into further details of the tests, it is evident that the new telescope will afford boundless possibilities for the study of the stellar universe.[*] The structure and extent of the galactic system, and the motions of the stars comprising it; the distribution, distances, and dimensions of the spiral nebul? their motions, rotation, and mode of development; the origin of the stars and the successive stages in their life history: these are some of the great questions which the new telescope must help to answer. In such an embarrassment of riches the chief difficulty is to withstand the temptation toward scattering of effort, and to form an observing programme directed toward the solution of crucial problems rather than the accumulation of vast stores of miscellaneous data. This programme will be supplemented by an extensive study of the sun, the only star near enough the earth to be examined in detail, and by a series of laboratory investigations involving the experimental imitation of solar and stellar conditions, thus aiding in the interpretation of celestial phenomena.

[Footnote *: It is not adapted for work on the sun, as the mirrors would be distorted by its heat. Three other telescopes, especially designed for solar observations, are in use on Mount Wilson.]

CHAPTER II

GIANT STARS

Our ancestral sun, as pictured by Laplace, originally extended in a state of luminous vapor beyond the boundaries of the solar system. Rotating upon its axis, it slowly contracted through loss of heat by radiation, leaving behind it portions of its mass, which condensed to form the planets. Still gaseous, though now denser than water, it continues to pour out the heat on which our existence depends, as it shrinks imperceptibly toward its ultimate condition of a cold and darkened globe.

Laplace's hypothesis has been subjected in recent years to much criticism, and there is good reason to doubt whether his description of the mode of evolution of our solar system is correct in every particular. All critics agree, however, that the sun was once enormously larger than it now is, and that the planets originally formed part of its distended mass.

Even in its present diminished state, the sun is huge beyond easy conception. Our own earth, though so minute a fragment of the primeval sun, is nevertheless so large that some parts of its surface have not yet been explored. Seen beside the sun, by an observer on one of the planets, the earth would appear as an insignificant speck, which could be swallowed with ease by the whirling vortex of a sun-spot. If the sun were hollow, with the earth at its centre, the moon, though 240,000 miles from us, would have room and to spare in which to describe its orbit, for the sun is 865,000 miles in diameter, so that its volume is more than a million times that of the earth.

But what of the stars, proved by the spectroscope to be self-luminous, intensely hot, and formed of the same chemical elements that constitute the sun and the earth? Are they comparable in size with the sun? Do they occur in all stages of development, from infancy to old age? And if such stages can be detected, do they afford indications of the gradual diminution in volume which Laplace imagined the sun to experience?

STAR IMAGES

Prior to the application of the powerful new engine of research described in this article we have had no means of measuring the diameters of the stars. We have measured their distances and their motions, determined their chemical composition, and obtained undeniable evidence of progressive development, but even in the most powerful telescopes their images are so minute that they appear as points rather than as disks. In fact, the larger the telescope and the more perfect the atmospheric conditions at the observer's command, the smaller do these images appear. On the photographic plate, it is true, the stars are recorded as measurable disks, but these are due to the spreading of the light from their bright point-like images, and their diameters increase as the exposure time is prolonged. From the images of the brighter stars rays of light project in straight lines, but these also are instrumental phenomena, due to diffraction of light by the steel bars that support the small mirror in the tube of reflecting telescopes. In a word, the stars are so remote that the largest and most perfect telescopes show them only as extremely minute needle-points of light, without any trace of their true disks.

How, then, may we hope to measure their diameters? By using, as the man of science must so often do, indirect means when the direct attack fails. Most of the remarkable progress of astronomy during the last quarter-century has resulted from the application of new and ingenious devices borrowed from the physicist. These have multiplied to such a degree that some of our observatories are literally physical laboratories, in which the sun and stars are examined by powerful spectroscopes and other optical instruments that have recently advanced our knowledge of physics by leaps and bounds. In the present case we are indebted for our star-measuring device to the distinguished physicist Professor Albert A. Michelson, who has contributed a long array of novel apparatus and methods to physics and astronomy.

THE INTERFEROMETER

The instrument in question, known as the interferometer, had previously yielded a remarkable series of results when applied in its various forms to the solution of fundamental problems. To mention only a few of those that have helped to establish Michelson's fame, we may recall that our exact knowledge of the length of the international metre at Sevres, the world's standard of measurement, was obtained by him with an interferometer in terms of the invariable length of light-waves. A different form of interferometer has more recently enabled him to measure the minute tides within the solid body of the earth--not the great tides of the ocean, but the slight deformations of the earth's body, which is as rigid as steel, that are caused by the varying attractions of the sun and moon. Finally, to mention only one more case, it was the Michelson-Morley experiment, made years ago with still another form of interferometer, that yielded the basic idea from which the theory of relativity was developed by Lorentz and Einstein.

The history of the method of measuring star diameters is a very curious one, showing how the most promising opportunities for scientific progress may lie unused for decades. The fundamental principle of the device was first suggested by the great French physicist Fizeau in 1868. In 1874 the theory was developed by the French astronomer Stephan, who observed interference fringes given by a large number of stars, and rightly concluded that their angular diameters must be much smaller than 0.158 of a second of arc, the smallest measurable with his instrument. In 1890 Michelson, unaware of the earlier work, published in the Philosophical Magazine a complete description of an interferometer capable of determining with surprising accuracy the distance between the components of double stars so close together that no telescope can separate them. He also showed how the same principle could be applied to the measurement of star diameters if a sufficiently large interferometer could be built for this purpose, and developed the theory much more completely than Stephan had done. A year later he measured the diameters of Jupiter's satellites by this means at the Lick Observatory. But nearly thirty years elapsed before the next step was taken. Two causes have doubtless contributed to this delay. Both theory and experiment have demonstrated the extreme sensitiveness of the "interference fringes," on the observation of which the method depends, and it was generally supposed by astronomers that disturbances in the earth's atmosphere would prevent them from being clearly seen with large telescopes. Furthermore, a very large interferometer, too large to be carried by any existing telescope, was required for the star-diameter work, though close double stars could have been easily studied by this device with several of the large telescopes of the early nineties. But whatever the reasons, a powerful method of research lay unused.

The approaching completion of the 100-inch telescope of the Mount Wilson Observatory led me to suggest to Professor Michelson, before the United States entered the war, that the method be thoroughly tested under the favorable atmospheric conditions of Southern California. He was at that time at work on a special form of interferometer, designed to determine whether atmospheric disturbances could be disregarded in planning large-scale experiments. But the war intervened, and all of our efforts were concentrated for two years on the solution of war problems.[*] In 1919, as soon as the 100-inch telescope had been completed and tested, the work was resumed on Mount Wilson.

[Footnote *: Professor Michelson's most important contribution during the war period was a new and very efficient form of range-finder, adopted for use by the U. S. Navy.]

A LABORATORY EXPERIMENT

The principle of the method can be most readily seen by the aid of an experiment which any one can easily perform for himself with simple apparatus. Make a narrow slit, a few thousandths of an inch in width, in a sheet of black paper, and support it vertically before a brilliant source of light. Observe this from a distance of 40 or 50 feet with a small telescope magnifying about 30 diameters. The object-glass of the telescope should be covered with an opaque cap, pierced by two circular holes about one-eighth of an inch in diameter and half an inch apart. The holes should be on opposite sides of the centre of the object-glass and equidistant from it, and the line joining the holes should be horizontal. When this cap is removed the slit appears as a narrow vertical band with much fainter bands on both sides of it. With the cap in place, the central bright band appears to be ruled with narrow vertical lines or fringes produced by the "interference"[*] of the two pencils of light coming through different parts of the object-glass from the distant slit. Cover one of the holes, and the fringes instantly disappear. Their production requires the joint effect of the two light-pencils.

[Footnote *: For an explanation of the phenomena of interference, see any encyclop鐯dia or book on physics.]

Now suppose the two holes over the object-glass to be in movable plates, so that their distance apart can be varied. As they are gradually separated the narrow vertical fringes become less and less distinct, and finally vanish completely. Measure the distance between the holes and divide this by the wavelength of light, which we may call 1/50000 of an inch. The result is the angular width of the distant slit. Knowing the distance of the slit, we can at once calculate its linear width. If for the slit we substitute a minute circular hole, the method of measurement remains the same, but the angular diameter as calculated above must be multiplied by 1.22.[*]

[Footnote *: More complete details may be found in Michelson's Lowell Lectures on "Light-Waves and Their Uses," University of Chicago Press, 1907.]

To measure the diameter of a star we proceed in a similar way, but, as the angle it subtends is so small, we must use a very large telescope, for the smaller the angle the farther apart must be the two holes over the object-glass (or the mirror, in case a reflecting telescope is employed). In fact, when the holes are moved apart to the full aperture of the 100-inch Hooker telescope, the interference fringes are still visible even with the star Betelgeuse, though its angular diameter is perhaps as great as that of any other star. Thus, we must build an attachment for the telescope, so arranged as to permit us to move the openings still farther apart.

THE 20-FOOT INSTRUMENT

The 20-foot interferometer designed by Messrs. Michelson and Pease, and constructed in the Mount Wilson Observatory instrument-shop, is shown in the diagram (Fig. 23) and in a photograph of the upper end of the skeleton tube of the telescope (Fig. 24). The light from the star is received by two flat mirrors (MI, M4) which project beyond the tube and can be moved apart along the supporting arm. These take the place of the two holes over the object-glass in our experiment. From these mirrors the light is reflected to a second pair of flat mirrors (M2, M3), which send it toward the 100-inch concave mirror (M5) at the bottom of the telescope tube. After this the course of the light is exactly as it would be if the mirrors M2, M3 were replaced by two holes over the 100-inch mirror. It is reflected to the convex mirror (M6), then back in a less rapidly convergent beam toward the large mirror. Before reaching it the light is caught by the plane mirror (M7) and reflected through an opening at the side of the telescope tube to the eye-piece E. Here the fringes are observed with a magnification ranging from 1,500 to 3,000 diameters.

In the practical application of this method to the measurement of star diameters, the chief problem was whether the atmosphere would be quiet enough to permit sharp interference fringes to be produced with light-pencils more than 100 inches apart. After successful preliminary tests with the 40-inch refracting telescope of the Yerkes Observatory, Professor Michelson made the first attempt to see the fringes with the 60-inch and 100-inch reflectors on Mount Wilson in September, 1919. He was surprised and delighted to find that the fringes were perfectly sharp and distinct with the full aperture of both these instruments. Doctor Anderson, of the observatory staff, then devised a special form of interferometer for the measurement of close double stars, and applied it with the 100-inch telescope to the measurement of the orbital motion of the close components of Capella, with results of extraordinary accuracy, far beyond anything attainable by previous methods. The success of this work strongly encouraged the more ambitious project of measuring the diameter of a star, and the 20-foot interferometer was built for this purpose.

The difficult and delicate problem of adjusting the mirrors of this instrument with the necessary extreme accuracy was solved by Professor Michelson during his visit to Mount Wilson in the summer of 1920, and with the assistance of Mr. Pease, of the observatory staff, interference fringes were observed in the case of certain stars when the mirrors were as much as 18 feet apart. All was thus in readiness for a decisive test as soon as a suitable star presented itself.

THE GIANT BETELGEUSE

Russell, Shapley, and Eddington had pointed out Betelgeuse (Arabic for "the giant's shoulder"), the bright red star in the constellation of Orion (Fig. 25), as the most favorable of all stars for measurement, and the last-named had given its angular diameter as 0.051 of a second of arc. This deduction from theory appeared in his recent presidential address before the British Association for the Advancement of Science, in which Professor Eddington remarked: "Probably the greatest need of stellar astronomy at the present day, in order to make sure that our theoretical deductions are starting on the right lines, is some means of measuring the apparent angular diameter of stars." He then referred to the work already in progress on Mount Wilson, but anticipated "that atmospheric disturbance will ultimately set the limit to what can be accomplished."

Measures with the interferometer show its angular diameter to be 0.047 of a second of arc, corresponding to a linear diameter of 215,000,000 miles, if the best available determination of its distance can be relied upon. This determination shows Betelgeuse to be 160 light-years from the earth. Light travels at the rate of 186,000 miles per second, and yet spends 160 years on its journey to us from this star.]

On December 13, 1920, Mr. Pease successfully measured the diameter of Betelgeuse with the 20-foot interferometer. As the outer mirrors were separated the interference fringes gradually became less distinct, as theory requires, and as Doctor Merrill had previously seen when observing Betelgeuse with the interferometer used for Capella. At a separation of 10 feet the fringes disappeared completely, giving the data required for calculating the diameter of the star. To test the perfection of the adjustment, the telescope was turned to other stars, of smaller angular diameter, which showed the fringes with perfect clearness. Turning back to Betelgeuse, they were seen beyond doubt to be absent. Assuming the mean wave-length of the light of this star to be 5750/10000000 of a millimetre, its angular diameter comes out 0.047 of a second of arc, thus falling between the values- -0.051 and 0.031 of a second--predicted by Eddington and Russell from slightly different assumptions. Subsequent corrections and repeated measurement will change Mr. Pease's result somewhat, but it is almost certainly within 10 or 15 per cent of the truth. We may therefore conclude that the angular diameter of Betelgeuse is very nearly the same as that of a ball one inch in diameter, seen at a distance of seventy miles.

Its angular diameter, measured at Mount Wilson by Pease with the 20-foot Michelson interferometer on April 15, 1921, is 0.022 of a second, in close agreement with Russell's predicted value of 0.019 of a second. The mean parallax of Arcturus, based upon several determinations, is 0.095 of a second, corresponding to a distance of 34 light-years. The linear diameter, computed from Pease's measure and this value of the distance is about 21 million miles.]

But this represents only the angle subtended by the star's disk. To learn its linear diameter, we must know its distance. Four determinations of the parallax, which determines the distance, have been made. Elkin, with the Yale heliometer, obtained 0.032 of a second of arc. Schlesinger, from photographs taken with the 30-inch Allegheny refractor, derived 0.016. Adams, by his spectroscopic method applied with the 60-inch Mount Wilson reflector, obtained 0.012. Lee's recent value, secured photographically with the 40-inch Yerkes refractor, is 0.022. The heliometer parallax is doubtless less reliable than the photographic ones, and Doctor Adams states that the spectral type and luminosity of Betelgeuse make his value less certain than in the case of most other stars. If we take a (weighted) mean value of 0.020 of a second, we shall probably not be far from the truth. This parallax represents the angle subtended by the radius of the earth's orbit (93,000,000 miles) at the distance of Betelgeuse. By comparing it with 0.047, the angular diameter of the star, we see that the linear diameter is about two and one-third times as great as the distance from the earth to the sun, or approximately 215,000,000 miles. Thus, if this measure of its distance is not considerably in error, Betelgeuse would nearly fill the orbit of Mars. All methods of determining the distances of the stars are subject to uncertainty, however, and subsequent measures may reduce this figure very appreciably. But there can be no doubt that the diameter of Betelgeuse exceeds 100,000,000 miles, and it is probably much greater.

The extremely small angle subtended by this enormous disk is explained by the great distance of the star, which is about 160 light-years. That is to say, light travelling at the rate of 186,000 miles per second spends 160 years in crossing the space that lies between us and Betelgeuse, whose tremendous proportions therefore seem so minute even in the most powerful telescopes.

STELLAR EVOLUTION

This actual measure of the diameter of Betelgeuse supplies a new and striking test of Russell's and Hertzsprung's theory of dwarf and giant stars. Just before the war Russell showed that our old methods of classifying the stars according to their spectra must be radically changed. Stars in an early stage of their life history may be regarded as diffuse gaseous masses, enormously larger than our sun, and at a much lower temperature. Their density must be very low, and their state that of a perfect gas. These are the "giants." In the slow process of time they contract through constant loss of heat by radiation. But, despite this loss, the heat produced by contraction and from other sources (see p. 82) causes their temperature to rise, while their color changes from red to bluish white. The process of shrinkage and rise of temperature goes on so long as they remain in the state of a perfect gas. But as soon as contraction has increased the density of the gas beyond a certain point the cycle reverses and the temperature begins to fall. The bluish-white light of the star turns yellowish, and we enter the dwarf stage, of which our own sun is a representative. The density increases, surpassing that of water in the case of the sun, and going far beyond this point in later stages. In the lapse of millions of years a reddish hue appears, finally turning to deep red. The falling temperature permits the chemical elements, existing in a gaseous state in the outer atmosphere of the star, to unite into compounds, which are rendered conspicuous by their characteristic bands in the spectrum. Finally comes extinction of light, as the star approaches its ultimate state of a cold and solid globe.

The distance of Antares, though not very accurately known, is probably not far from 350 light-years. Its angular diameter of 0.040 of a second would thus correspond to a linear diameter of about 400 million miles.]

We may thus form a new picture of the two branches of the temperature curve, long since suggested by Lockyer, on very different grounds, as the outline of stellar life. On the ascending side are the giants, of vast dimensions and more diffuse than the air we breathe. There are good reasons for believing that the mass of Betelgeuse cannot be more than ten times that of the sun, while its volume is at least a million times as great and may exceed eight million times the sun's volume. Therefore, its average density must be like that of an attenuated gas in an electric vacuum tube. Three-quarters of the naked-eye stars are in the giant stage, which comprises such familiar objects as Betelgeuse, Antares, and Aldebaran, but most of them are much denser than these greatly inflated bodies. The pinnacle is reached in the intensely hot white stars of the helium class, in whose spectra the lines of this gas are very conspicuous. The density of these stars is perhaps one-tenth that of the sun. Sirius, also very hot, is nearly twice as dense. Then comes the cooling stage, characterized, as already remarked, by increasing density, and also by increasing chemical complexity resulting from falling temperature. This life cycle is probably not followed by all stars, but it may hold true for millions of them.

The existence of giant and dwarf stars has been fully proved by the remarkable work of Adams and his associates on Mount Wilson, where his method of determining a star's distance and intrinsic luminosity by spectroscopic observations has already been applied to 2,000 stars. Discussion of the results leads at once to the recognition of the two great classes of giants and dwarfs. Now comes the work of Michelson and Pease to cap the climax, giving us the actual diameter of a typical giant star, in close agreement with predictions based upon theory. From this diameter we may conclude that the density of Betelgeuse is extremely low, in harmony with Russell's theory, which is further supported by spectroscopic analysis of the star's light, revealing evidence of the comparatively low temperature called for by the theory at this early stage of stellar existence.

TWO OTHER GIANTS

The diameter of Arcturus was successfully measured by Mr. Pease at Mount Wilson on April 15. As the mirrors of the interferometer were moved apart, the fringes gradually decreased in visibility until they finally disappeared at a mirror separation of 19.6 feet. Adopting a mean wave-length of 5600/10000000 of a millimetre for the light of Arcturus, this gives a value of 0.022 of a second of arc for the angular diameter of the star. If we use a mean value of 0.095 of a second for the parallax, the corresponding linear diameter comes out 21,000,000 miles. The angular diameter, as in the case of Betelgeuse, is in remarkably close agreement with the diameter predicted from theory. Antares, the third star measured by Mr. Pease, is the largest of all. If it is actually a member of the Scorpius-Centaurus group, as we have strong reason to believe, it is fully 350 light-years from the earth, and its diameter is about 400,000,000 miles.

Sun, diameter, 865,000 miles.

Arcturus, diameter, 21,000,000 miles.

Betelgeuse, diameter, 215,000,000 miles.

Antares, diameter, 400,000,000 miles.]

It now remains to make further measures of Betelgeuse, especially because its marked changes in brightness suggest possible variations in diameter. We must also apply the interferometer method to stars of the various spectral types, in order to afford a sure basis for future studies of stellar evolution. Unfortunately, only a few giant stars are certain to fall within the range of our present instrument. An interferometer of 70-feet aperture would be needed to measure Sirius accurately, and one of twice this size to deal with less brilliant white stars. A 100-foot instrument, if feasible to build, would permit objects representing most of the chief stages of stellar development to be measured, thus contributing in the highest degree to the progress of our knowledge of the life history of the stars. Fortunately, though the mechanical difficulties are great, the optical problem is insignificant, and the cost of the entire apparatus, though necessarily high, would be only a small fraction of that of a telescope of corresponding aperture, if such could be built. A 100-foot interferometer might be designed in many different forms, and one of these may ultimately be found to be within the range of possibility. Meanwhile the 20-foot interferometer has been improved so materially that it now promises to yield approximate measures of stars at first supposed to be beyond its capacity.

Like Betelgeuse and Antares, it is notable for its red color, which accounts for the fact that its image on this photograph is hardly more conspicuous than the images of stars which are actually much fainter but contain a larger proportion of blue light, to which the photographic plates here employed are more sensitive than to red or yellow. Aldebaran is about 50 light-years from the earth. Interferometer measures, now in progress on Mount Wilson, indicate that its angular diameter is about 0.020 of a second.]

While the theory of dwarf and giant stars and the measurements just described afford no direct evidence bearing on Laplace's explanation of the formation of planets, they show that stars exist which are comparable in diameter with our solar system, and suggest that the sun must have shrunk from vast dimensions. The mode of formation of systems like our own, and of other systems numerously illustrated in the heavens, is one of the most fascinating problems of astronomy. Much light has been thrown on it by recent investigations, rendered possible by the development of new and powerful instruments and by advances in physics of the most fundamental character. All the evidence confirms the existence of dwarf and giant stars, but much work must be done before the entire course of stellar evolution can be explained.

CHAPTER III

COSMIC CRUCIBLES

"Shelter during Raids," marking the entrance to underground passages, was a sign of common occurrence and sinister suggestion throughout London during the war. With characteristic ingenuity and craftiness, ostensibly for purposes of peace but with bomb-carrying capacity as a prime specification, the Zeppelin had been developed by the Germans to a point where it seriously threatened both London and Paris. Searchlights, range-finders, and anti-aircraft guns, surpassed by the daring ventures of British and French airmen, would have served but little against the night invader except for its one fatal defect--the inflammable nature of the hydrogen gas that kept it aloft. A single explosive bullet served to transform a Zeppelin into a heap of scorched and twisted metal. This characteristic of hydrogen caused the failure of the Zeppelin raids.

Had the war lasted a few months longer, however, the work of American scientists would have made our counter-attack in the air a formidable one. At the signing of the armistice hundreds of cylinders of compressed helium lay at the docks ready for shipment abroad. Extracted from the natural gas of Texas wells by new and ingenious processes, this substitute for hydrogen, almost as light and absolutely uninflammable, produced in quantities of millions of cubic feet, would have made the dirigibles of the Allies masters of the air. The special properties of this remarkable gas, previously obtainable only in minute quantities, would have sufficed to reverse the situation.

SOLAR HELIUM

Helium, as its name implies, is of solar origin. In 1868, when Lockyer first directed his spectroscope to the great flames or prominences that rise thousands of miles, sometimes hundreds of thousands, above the surface of the sun, he instantly identified the characteristic red and blue radiations of hydrogen. In the yellow, close to the position of the well-known double line of sodium, but not quite coincident with it, he detected a new line, of great brilliancy, extending to the highest levels. Its similarity in this respect with the lines of hydrogen led him to recognize the existence of a new and very light gas, unknown to terrestrial chemistry.

Many years passed before any chemical laboratory on earth was able to match this product of the great laboratory of the sun. In 1896 Ramsay at last succeeded in separating helium, recognized by the same yellow line in its spectrum, in minute quantities from the mineral uraninite. Once available for study under electrical excitation in vacuum tubes, helium was found to have many other lines in its spectrum, which have been identified in the spectra of solar prominences, gaseous nebul? and hot stars. Indeed, there is a stellar class known as helium stars, because of the dominance of this gas in their atmospheres.

In these luminous gaseous clouds, which sometimes rise to elevations exceeding half the sun's diameter, the new gas helium was discovered by Lockyer in 1868. Helium was not found on the earth until 1896. Since then it has been shown to be a prominent constituent of nebula and hot stars.

The chief importance of helium lies in the clue it has afforded to the constitution of matter and the transmutation of the elements. Radium and other radioactive substances, such as uranium, spontaneously emit negatively charged particles of extremely small mass (electrons), and also positively charged particles of much greater mass, known as alpha particles. Rutherford and Geiger actually succeeded in counting the number of alpha particles emitted per second by a known mass of radium, and showed that these were charged helium atoms.

To discuss more at length the extraordinary characteristics of helium, which plays so large a part in celestial affairs, would take us too far afield. Let us therefore pass to another case in which a fundamental discovery, this time in physics, was first foreshadowed by astronomical observation.

SUN-SPOTS AS MAGNETS

No archeologist, whether Young or Champollion deciphering the Rosetta Stone, or Rawlinson copying the cuneiform inscription on the cliff of Behistun, was ever faced by a more fascinating problem than that which confronts the solar physicist engaged in the interpretation of the hieroglyphic lines of sun-spot spectra. The colossal whirling storms that constitute sun-spots, so vast that the earth would make but a moment's scant mouthful for them, differ materially from the general light of the sun when examined with the spectroscope. Observing them visually many years ago, the late Professor Young, of Princeton, found among their complex features a number of double lines which he naturally attributed, in harmony with the physical knowledge of the time, to the effect of "reversal" by superposed layers of vapors of different density and temperature. What he actually saw, however, as was proved at the Mount Wilson Observatory in 1908, was the effect of a powerful magnetic field on radiation, now known as the Zeeman effect.

An image of the sun about 16 inches in diameter is formed in the laboratory at the base of the tower. Below this, in a well extending 80 feet into the earth, is the powerful spectroscope with which the magnetic fields in sun-spots and the general magnetic field of the sun are studied.]

Faraday was the first to detect the influence of magnetism on light. Between the poles of a large electromagnet, powerful for those days (1845), he placed a block of very dense glass. The plane of polarization of a beam of light, which passed unaffected through the glass before the switch was closed, was seen to rotate when the magnetic field was produced by the flow of the current. A similar rotation is now familiar in the well-known tests of sugars--lulose and dextrose--which rotate plane-polarized light to left and right, respectively.

But in this first discovery of a relationship between light and magnetism Faraday had not taken the more important step that he coveted--to determine whether the vibration period of a light-emitting particle is subject to change in a magnetic field. He attempted this in 1862--the last experiment of his life. A sodium flame was placed between the poles of a magnet, and the yellow lines were watched in a spectroscope when the magnet was excited. No change could be detected, and none was found by subsequent investigators until Zeeman, of Leiden, with more powerful instruments made his famous discovery, the twenty-fifth anniversary of which has recently been celebrated.

Showing the large magnet (on the left) and the spectroscopes used for the study of the effect of magnetism on radiation. A single line in the spectrum is split by the magnetic field into from three to twenty-one components, as illustrated in Fig. 34. The corresponding lines in the spectra of sun-spots are split up in precisely the same way, thus indicating the presence of powerful magnetic fields in the sun.]

His method of procedure was similar to Faraday's, but his magnet and spectroscope were much more powerful, and a theory due to Lorentz, predicting the nature of the change to be expected, was available as a check on his results. When the current was applied the lines were seen to widen. In a still more powerful magnetic field each of them split into two components (when the observation was made along the lines of force), and the light of the components of each line was found to be circularly polarized in opposite directions. Strictly in harmony with Lorentz's theory, this splitting and polarization proved the presence in the luminous vapor of exactly such negatively charged electrons as had been indicated there previously by very different experimental methods.

In 1908 great cyclonic storms, or vortices, were discovered at the Mount Wilson Observatory centring in sun-spots. Such whirling masses of hot vapors, inferred from Sir Joseph Thomson's results to contain electrically charged particles, should give rise to a magnetic field. This hypothesis at once suggested that the double lines observed by Young might really represent the Zeeman effect. The test was made, and all the characteristic phenomena of radiation in a magnetic field were found.

Thus a great physical experiment is constantly being performed for us in the sun. Every large sunspot contains a magnetic field covering many thousands of square miles, within which the spectrum lines of iron, manganese, chromium, titanium, vanadium, calcium, and other metallic vapors are so powerfully affected that their widening and splitting can be seen with telescopes and spectroscopes of moderate size.

THE TOWER TELESCOPE

Both of these illustrations show how the physicist and chemist, when adequately armed for astronomical attack, can take advantage in their studies of the stupendous processes visible in cosmic crucibles, heated to high temperatures and influenced, as in the case of sun-spots, by intense magnetic fields. Certain modern instruments, like the 60-foot and 150-foot tower telescopes on Mount Wilson, are especially designed for observing the course of these experiments. The second of these telescopes produces at a fixed point in a laboratory an image of the sun about 16 inches in diameter, thus enlarging the sun-spots to such a scale that the magnetic phenomena of their various parts can be separately studied. This analysis is accomplished with a spectroscope 80 feet in length, mounted in a subterranean chamber beneath the tower. The varied results of such investigations cannot be described here. Only one of them may be mentioned--the discovery that the entire sun, rotating on its axis, is a great magnet. Hence we may reasonably infer that every star, and probably every planet, is also a magnet, as the earth has been known to be since the days of Gilbert's "De Magnete." Here lies one of the best clues for the physicist who seeks the cause of magnetism, and attempts to produce it, as Barnett has recently succeeded in doing, by rapidly whirling masses of metal in the laboratory.

Photographed with the spectroheliograph. The electric vortex that causes the magnetic field of the spot lies at a lower level, and is not shown by such photographs.]

Perhaps a word of caution should be interpolated at this point. Solar magnetism in no wise accounts for the sun's gravitational power. Indeed, its attraction cannot be felt by the most delicate instruments at the distance of the earth, and would still be unknown were it not for the influence of magnetism on light.

Auroras, magnetic storms, and such electric currents as those that recently deranged several Atlantic cables are due, not to the magnetism of the sun or its spots, but probably to streams of electrons, shot out from highly disturbed areas of the solar surface surrounding great sun-spots, traversing ninety-three million miles of the ether of space, and penetrating deep into the earth's atmosphere. These striking phenomena lead us into another chapter of physics, which limitations of space forbid us to pursue.

STELLAR CHEMISTRY

Let us turn again to chemistry, and see where experiments performed in cosmic laboratories can serve as a guide to the investigator. A spinning solar tornado, incomparably greater in scale than the devastating whirlwinds that so often cut narrow paths of destruction through town and country in the Middle West, gradually gives rise to a sun-spot. The expansion produced by the centrifugal force at the centre of the storm cools the intensely hot gases of the solar atmosphere to a point where chemical union can occur. Titanium and oxygen, too hot to combine in most regions of the sun, join to form the vapor of titanium oxide, characterized in the sunspot spectrum by fluted bands, made up of hundreds of regularly spaced lines. Similarly magnesium and hydrogen combine as magnesium hydride and calcium and hydrogen form calcium hydride. None of these compounds, stable at the high temperatures of sun-spots, has been much studied in the laboratory. The regions in which they exist, though cooler than the general atmosphere of the sun, are at temperatures of several thousand degrees, attained in our laboratories only with the aid of such devices as powerful electric furnaces.

The upper and lower strips show lines in the spectrum of chromium, observed without a magnetic field. When subjected to the influence of magnetism, these single lines are split into several components. Thus the first line on the right is resolved by the field into three components, one of which (plane polarized) appears in the second strip, while the other two, which are polarized in a plane at right angles to that of the middle component, are shown on the third strip. The next line is split by the magnetic field into twelve components, four of which appear in the second strip and eight in the third. The magnetic fields in sun-spots affect these lines in precisely the same way.]

It is interesting to follow our line of reasoning to the stars, which differ widely in temperature at various stages in their life-cycle.[*] A sun-spot is a solar tornado, wherein the intensely hot solar vapors are cooled by expansion, giving rise to the compounds already named. A red star, in Russell's scheme of stellar evolution, is a cooler sun, vast in volume and far more tenuous than atmospheric air when in the initial period of the "giant" stage, but compressed and denser than water in the "dwarf" stage, into which our sun has already entered as it gradually approaches the last phases of its existence. Therefore we should find, throughout the entire atmosphere of such stars, some of the same compounds that are produced within the comparatively small limits of a sun-spot. This, of course, on the correct assumption that sun and stars are made of the same substances. Fowler has already identified the bands of titanium oxide in such red stars as the giant Betelgeuse, and in others of its class. It is safe to predict that an interesting chapter in the chemistry of the future will be based upon the study of such compounds, both in the laboratory and under the progressive temperature conditions afforded by the countless stellar "giants" and "dwarfs" that precede and follow the solar state.

[Footnote *: See Chapter II.]

ASTROPHYSICAL LABORATORIES

It is precisely in this long sequence of physical and chemical changes that the astrophysicist and the astrochemist can find the means of pushing home their attack. It is true, of course, that the laboratory investigator has a great advantage in his ability to control his experiments, and to vary their progress at will. But by judicious use of the transcendental temperatures, far out ranging those of his furnaces, and extreme conditions, which he can only partially imitate, afforded by the sun, stars, and nebula he may greatly widen the range of his inquiries. The sequence of phenomena seen during the growth of a sun-spot, or the observation of spots of different sizes, and the long series of successive steps that mark the rise and decay of stellar life, resemble the changes that the experimenter brings about as he increases and diminishes the current in the coils of his magnet or raises and lowers the temperature of his electric furnace, examining from time to time the spectrum of the glowing vapors, and noting the changes shown by the varying appearance of their lines.

The upper spectrum is that of titanium in the flame of the electric arc, where its combination with oxygen gives rise to the bands of titanium oxide (Fowler). The lower strip shows the spectrum of the red star Mira (Omicron Ceti), as drawn by Cortie at Stonyhurst. The bands of titanium oxide are clearly present in the star.]

The upper strip shows a portion of the spectrum of a sun-spot (Ellerman); the lower one the corresponding region of the spectrum of titanium oxide (King). The fluted bands of the oxide spectrum are easily identified in the spot, where they indicate that titanium and oxygen, too hot to combine in the solar atmosphere, unite in the spot because of the cooling produced by expansion in the vortex.]

Astronomical observations of this character, it should be noted, are most effective when constantly tested and interpreted by laboratory experiment. Indeed, a modern astrophysical observatory should be equipped like a great physical laboratory, provided on the one hand with telescopes and accessory apparatus of the greatest attainable power, and on the other with every device known to the investigator of radiation and the related physical and chemical phenomena. Its telescopes, especially designed with the aims of the physicist and chemist in view, bring images of sun, stars, nebula and other heavenly bodies within the reach of powerful spectroscopes, sensitive bolometers and thermopiles, and the long array of other appliances available for the measurement and analysis of radiation. Its electric furnaces, arcs, sparks, and vacuum tubes, its apparatus for increasing and decreasing pressure, varying chemical conditions, and subjecting luminous gases and vapors to the influence of electric and magnetic fields, provide the means of imitating celestial phenomena, and of repeating and interpreting the experiments observed at the telescope. And the advantage thus derived, as we have seen, is not confined to the astronomer, who has often been able, by making fundamental physical and chemical discoveries, to repay his debt to the physicist and chemist for the apparatus and methods which he owes to them.

NEWTON AND EINSTEIN

Take, for another example, the greatest law of physics--Newton's law of gravitation. Huge balls of lead, as used by Cavendish, produce by their gravitational effect a minute rotation of a delicately suspended bar, carrying smaller balls at its extremities. But no such feeble means sufficed for Newton's purpose. To prove the law of gravitation he had recourse to the tremendous pull on the moon of the entire mass of the earth, and then extended his researches to the mutual attractions of all the bodies of the solar system. Later Herschel applied this law to the suns which constitute double stars, and to-day Adams observes from Mount Wilson stars falling with great velocity toward the centre of the galactic system under the combined pull of the millions of objects that compose it. Thus full advantage has been taken of the possibility of utilizing the great masses of the heavenly bodies for the discovery and application of a law of physics and its reciprocal use in explaining celestial motions.

Two lead balls, each two inches in diameter, are attached to the ends of a torsion rod six feet long, which is suspended by a fine wire. The experiment consists in measuring the rotation of the suspended system, caused by the gravitational attraction of two lead spheres, each twelve inches in diameter, acting on the two small lead balls.]

Or consider the Einstein theory of relativity, the truth or falsity of which is no less fundamental to physics. Its inception sprang from the Michelson-Morley experiment, made in a laboratory in Cleveland, which showed that motion of the earth through the ether of space could not be detected. All of the three chief tests of Einstein's general theory are astronomical--because of the great masses required to produce the minute effects predicted: the motion of the perihelion of Mercury, the deflection of the light of a star by the attraction of the sun, and the shift of the lines of the solar spectrum toward the red--questions not yet completely answered.

But it is in the study of the constitution of matter and the evolution of the elements, the deepest and most critical problem of physics and chemistry, that the extremes of pressure and temperature in the heavenly bodies, and the prevalence of other physical conditions not yet successfully imitated on earth, promise the greatest progress. It fortunately happens that astrophysical research is now at the very apex of its development, founded as it is upon many centuries of astronomical investigation, rejuvenated by the introduction into the observatory of all the modern devices of the physicist, and strengthened with instruments of truly extraordinary range and power. These instruments bring within reach experiments that are in progress on some minute region of the sun's disk, or in some star too distant even to be glimpsed with ordinary telescopes. Indeed, the huge astronomical lenses and mirrors now available serve for these remote light-sources exactly the purpose of the lens or mirror employed by the physicist to project upon the slit of his spectroscope the image of a spark or arc or vacuum tube within which atoms and molecules are exposed to the influence of the electric discharge. The physicist has the advantage of complete control over the experimental conditions, while the astrophysicist must observe and interpret the experiments performed for him in remote laboratories. In actual practice, the two classes of work must be done in the closest conjunction, if adequate utilization is to be made of either. And this is only natural, for the trend of recent research has made clear the fact that one of the three greatest problems of modern astronomy and astrophysics, ranking with the structure of the universe and the evolution of celestial bodies, is the constitution of matter. Let us see why this is so.

TRANSMUTATION OF THE ELEMENTS

The dream of the alchemist was to transmute one element into another, with the prime object of producing gold. Such transmutation has been actually accomplished within the last few years, but the process is invariably one of disintegration--the more complex elements being broken up into simpler constituents. Much remains to be done in this same direction; and here the stars and nebula which show the spectra of the elements under a great variety of conditions, should help to point the way. The progressive changes in spectra, from the exclusive indications of the simple elements hydrogen, helium, nitrogen, possibly carbon, and the terrestrially unknown gas nebulium in the gaseous nebula to the long list of familiar substances, including several chemical compounds, in the red stars, may prove to be fundamentally significant when adequately studied from the standpoint of the investigator of atomic structure. The existing evidence seems to favor the view, recently expressed by Saha, that many of these differences are due to varying degrees of ionization, the outer electrons of the atoms being split off by high temperature or electrical excitation. It is even possible that cosmic crucibles, unrivalled by terrestrial ones, may help materially to reveal the secret of the formation of complex elements from simpler ones. Physicists now believe that all of the elements are compounded of hydrogen atoms, bound together by negative electrons. Thus helium is made up of four hydrogen atoms, yet the atomic weight of helium (4) is less than four times that of hydrogen (1.008). The difference may represent the mass of the electrical energy released when the transmutation occurred.

The gas "nebulium," not yet found on the earth, is the most characteristic constituent of irregular nebula Nebulium is recognized by two green lines in its spectrum, which cause the green color of nebula of the gaseous type.]

Eddington has speculated in a most interesting way on this possible source of stellar heat in his recent presidential address before the British Association for the Advancement of Science (see Nature, September 2, 1920). He points out that the old contraction hypothesis, according to which the source of solar and stellar heat was supposed to reside in the slow condensation of a radiating mass of gas under the action of gravity, is wholly inadequate to explain the observed phenomena. If the old view were correct, the earlier history of a star, from the giant stage of a cool and diaphanous gas to the period of highest temperature, would be run through within eighty thousand years, whereas we have the best of evidence that many thousands of centuries would not suffice. Some other source of energy is imperatively needed. If 5 per cent of a star's mass consists originally of hydrogen atoms, which gradually combine in the slow process of time to form more complex elements, the total heat thus liberated would more than suffice to account for all demands, and it would be unnecessary to assume the existence of any other source of heat.

Luminous matter, in every variety of physical and chemical state, is available for study in the most diverse celestial objects, from the spiral and irregular nebul?through all the types of stars. Doctor van Maanen's measures of the Mount Wilson photographs indicate outward motion along the arms of spiral nebul? while the spectroscope shows them to be whirling at enormous velocities.]

COSMIC PRESSURES

This, it may fairly be said, is very speculative, but the fact remains that celestial bodies appear to be the only places in which the complex elements may be in actual process of formation from their known source--hydrogen. At least we may see what a vast variety of physical conditions these cosmic crucibles afford. At one end of the scale we have the excessively tenuous nebul? the luminosity of which, mysterious in its origin, resembles the electric glow in our vacuum tubes. Here we can detect only the lightest and simplest of the elements. In the giant stars, also extremely tenuous (the density of Betelgeuse can hardly exceed one-thousandth of an atmosphere) we observe the spectra of iron, manganese, titanium, calcium, chromium, magnesium, vanadium, and sodium, in addition to titanium oxide. The outer part of these bodies, from which light reaches us, must therefore be at a temperature of only a few thousand degrees, but vastly higher temperatures must prevail at their centres. In passing up the temperature curve more and more elements appear, the surface temperature rises, and the internal temperature may reach millions of degrees. At the same time the pressure within must also rise, reaching enormous figures in the last stages of stellar life. Cook has calculated that the pressure at the centre of the earth is between 4,000 and 10,000 tons per square inch, and this must be only a very small fraction of that attained within larger celestial bodies. Jeans has computed the pressure at the centre of two colliding stars as they strike and flatten, and finds it may be of the order of 1,000,000,000 tons per square inch--sufficient, if their diameter be equal to that of the sun--to vaporize them 100,000 times over.

Compare these pressures with the highest that can be produced on earth. If the German gun that bombarded Paris were loaded with a solid steel projectile of suitable dimensions, a muzzle velocity of 6,000 feet per second could be reached. Suppose this to be fired into a tapered hole in a great block of steel. The instantaneous pressure, according to Cook, would be about 7,000 tons per square inch, only 1/150000 of that possible through the collision of the largest stars.

Michelson is measuring the velocity of light between stations on Mount Wilson and Mount San Antonio. Astronomical observations afford the best means, however, of detecting any possible difference between the velocities of light of different colors. From studies of variable stars in the cluster Messier 5 Shapley concludes that if there is any difference between the velocities of blue and yellow light in free space it cannot exceed two inches in one second, the time in which light travels 186,000 miles.]

Finally, we may compare the effects of light pressure on the earth and stars. Twenty years ago Nichols and Hull succeeded, with the aid of the most sensitive apparatus, in measuring the minute displacements produced by the pressure of light. The effect is so slight, even with the brightest light-sources available, that great experimental skill is required to measure it. Yet in the case of some of the larger stars Eddington calculates that one-half of their mass is supported by radiation pressure, and this against their enormous gravitational attraction. In fact, if their mass were as great as ten times that of the sun, the radiation pressure would so nearly overcome the pull of gravitation that they would be likely to break up.

But enough has been said to illustrate the wide variety of experimental devices that stand at our service in the laboratories of the heavens. Here the physicist and chemist of the future will more and more frequently supplement their terrestrial apparatus, and find new clues to the complex problems which the amazing progress of recent years has already done so much to solve.

PRACTICAL VALUE OF RESEARCHES ON THE CONSTITUTION OF MATTER

The layman has no difficulty in recognizing the practical value of researches directed toward the improvement of the incandescent lamp or the increased efficiency of the telephone. He can see the results in the greatly decreased cost of electric illumination and the rapid extension of the range of the human voice. But the very men who have made these advances, those who have succeeded beyond all expectation in accomplishing the economic purposes in view, are most emphatic in their insistence upon the importance of research of a more fundamental character. Thus Vice-President J. J. Carty, of the American Telephone and Telegraph Company, who directs its great Department of Development and Research, and Doctor W. J. Whitney, Director of the Research Laboratory of the General Electric Company, have repeatedly expressed their indebtedness to the investigations of the physicist, made with no thought of immediate practical return. Faraday, studying the laws of electricity, discovered the principle which rendered the dynamo possible. Maxwell, Henry, and Hertz, equally unconcerned with material advantage, made wireless telegraphy practicable. In fact, all truly great advances are thus derived from fundamental science, and the future progress of the world will be largely dependent upon the provision made for scientific research, especially in the fields of physics and chemistry, which underlie all branches of engineering.

The constitution of matter, therefore, instead of appealing as a subject to research only to the natural philosopher or to the general student of science, is a question of the greatest practical concern. Already the by-products of investigations directed toward its elucidation have been numerous and useful in the highest degree. Helium has been already cited; X-rays hardly require mention; radium, which has so materially aided sufferers from cancer, is still better known. Wireless telephony and transcontinental telephony with wires were both rendered possible by studies of the nature of the electric discharge in vacuum tubes. Thus the "practical man," with his distrust of "pure" science, need not resent investments made for the purpose of advancing our knowledge of such fundamental subjects as physics and chemistry. On the contrary, if true to his name, he should help to multiply them many fold in the interest of economic and commercial development.

###

www.ingramcontent.com/pod-product-compliance
Lightning Source LLC
Chambersburg PA
CBHW070923180526
45168CB00005B/2129